AF575816

SPIDERS UP CLOSE

by Alan Walker

TABLE OF CONTENTS

A Crabtree Seedlings Book

Crabtree Publishing

crabtreebooks.com

2

Spiders Up Close

What do we really know about spiders?

Spiders are not insects. Insects have six legs and spiders have eight legs. Spiders are arachnids.

Spiders come in all shapes, sizes, and colors.

Spiders don't have ears, but the hairs on their legs help them detect sounds.

Most spiders have eight eyes.
Some have as many as twelve.
Some have no eyes at all!

The deadly black widow spider has eight eyes.

Spiders' legs help them move really fast.

Having eight legs
makes climbing easier.

A spider's fangs pierce its prey and inject venom.

Spiders have **fangs** for killing their **prey**.
prey

Spiders know how to hide. They blend into their surroundings.

To blend into your surroundings is called **camouflage**.

orb web

Some spiders make webs. They use the webs to catch their food.

Many strands in an orb web are sticky so that captured prey cannot escape.

trapdoor

Not all spiders make webs.
Some burrow into the ground.

The trapdoor spider
makes a burrow.
It makes a trapdoor
that it closes to hide.

No need to be afraid. Spiders can be helpful. They eat pests.

Many people are afraid of spiders, but most spiders are harmless.

Black Widow Spider
Red hourglass
shape on its
underside
Venomous to humans
Jumping Spider
Jumps to catch prey
Not venomous
to humans
Black and Yellow
Garden Spider
Large orb-weaving spider
Not venomous
to humans

Mexican Red-Knee Tarantula
Often sold as exotic pets
Not venomous to humans
Cellar Spider
Also called *Daddy Long Legs*
Not venomous to humans
Brown Recluse Spider
Has six eyes
Venomous to humans

Spider Body Parts

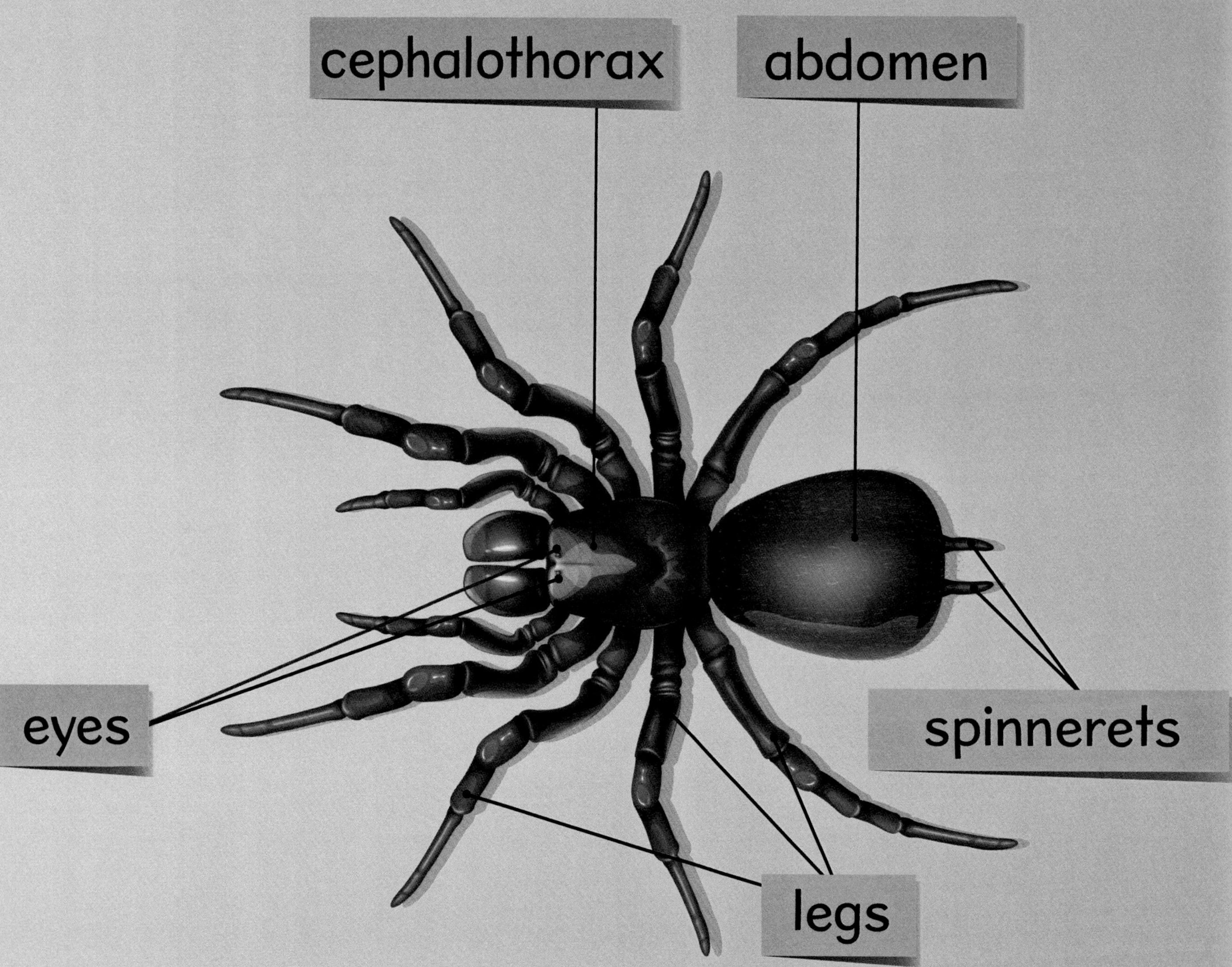

Glossary

arachnids (uh-RAK-nidz): Arachids are animals with eight legs, two body sections, and no wings or antennae.

camouflage (KAM-uh-flahzh): To camouflage is to hide by coloring or covering to look like the surroundings.

fangs (FANGZ): Fangs are long, pointed teeth.

pests (PESTS): Pests are insects that destroy or damage flowers, fruits, or vegetables.

prey (PRAY): Prey is an animal that is eaten by another animal.

venom (VEN-uhm): Venom is poison produced by some snakes and spiders. It is usually passed into a victim's body through a bite or sting.

Index

School-to-Home Support for Caregivers and Teachers

This book helps children grow by letting them practice reading. Here are a few guiding questions to help the reader build his or her comprehension skills. Possible answers appear here in red.

Before Reading

- **What do I think this book is about?** I think this book is about different kinds of spiders. I think this book is about how spiders can be helpful to people.
- **What do I want to learn about this topic?** I want to learn more about which spiders are venomous. I want to learn how a spider makes a web.

During Reading

- **I wonder why...** I wonder why spiders are not insects. I wonder why spiders have more than two eyes.
- **What have I learned so far?** I have learned that spiders have eight legs and are called arachnids. I have learned that not all spiders make webs to catch prey, and that some burrow into the ground to catch prey.

After Reading

- **What details did I learn about this topic?** I have learned that spiders can be helpful to people by eating insect pests. I have learned that most spiders are harmless to people.
- **Read the book again and look for the glossary words.** I see the word *fangs* on page 11, and the word *camouflage* on page 13. The other glossary words are found on page 23.

Crabtree Publishing

crabtreebooks.com 800-387-7650

In Canada: We acknowledge the financial support of the Government of Canada through the Canada Book Fund for our publishing activities.

Published in Canada
Crabtree Publishing
616 Welland Avenue
St. Catharines, Ontario
L2M 5V6

Published in the United States
Crabtree Publishing
347 Fifth Avenue
Suite 1402-145
New York, NY 10016

Written by: Alan Walker

Photo Credits: Cover © By Marek Velechovsky-Shutterstock.com www.shutterstock.com, www.istock.com. PGs 2-3; flukyfluky, Smitt. PGs 4-5; ConstantinCornel, AlinaMD, gutaper. PGs 6-7; ConstantinCornel, rticknor. PGs 8-9; davemhuntphotography, Dmitrii Erekhinskii. PGs 10-11; Peter Waters, jakkaphol phothongnak. PGs 12-13; dennisvdw, Luis Jimenez Benito, yod67. PGs14-15; LuckyToBeThere, Sergey_Ko. PGs 16-17; Federico.Crovetto. PGs 18-19; SHipskyy, _jure. PGs 20-21; feedough, JasonOndreicka, Macrolife.it, katoosha, Fotocute, N-sky, George Chernilevsky. PGs 22-23; BlueRingMedia.

Hardcover 978-1-0396-4465-6
Paperback 978-1-0396-4656-8

Printed in Canada/092023/CPC20230901

Library and Archives Canada Cataloguing in Publication
Available at the Library and Archives Canada

Library of Congress Cataloging-in-Publication Data
Available at the Library of Congress